Polar
dinosaurs
of Australia

Museum Victoria
Nature Series

Polar dinosaurs of Australia

Thomas H. Rich

MUSEUMVICTORIA

Polar dinosaurs of Australia

Published by
Museum Victoria

© Text copyright
Museum Victoria 2007

© Images copyright Museum
Victoria unless otherwise noted.
Museum Victoria has made every
effort to obtain copyright and
moral permission for use of all
images. Please advise us of any
errors or omissions.

Museum Victoria
GPO Box 666
Melbourne VIC 3001 Australia
Telephone: +61 3 8347 7777
www.museumvictoria.com.au

Dr J. Patrick Greene
CHIEF EXECUTIVE OFFICER

Dr Robin Hirst
DIRECTOR, COLLECTIONS,
RESEARCH AND EXHIBITIONS

Dr John Long
HEAD, SCIENCE DEPARTMENT

PRINTED BY
BPA Print Group

DESIGNED BY
Stephen Horsley at Propellant

National Library of Australia
Cataloguing-in-Publication data:
Rich, Thomas H. V.
Polar dinosaurs of Australia.

1st ed.
Bibliography.
ISBN 9780975837023.

1. Dinosaurs - Victoria.
2. Dinosaurs - Polar regions.
3. Paleontology - Victoria.
 I. Museum Victoria. II. Title.

567.909945

COVER IMAGE
Hypsilophodontids in the snow.
Adapted from a painting by
Peter Trusler.

PAGE ii
Dinosaur Cove, the Otway
Ranges, Victoria.

CONTENTS

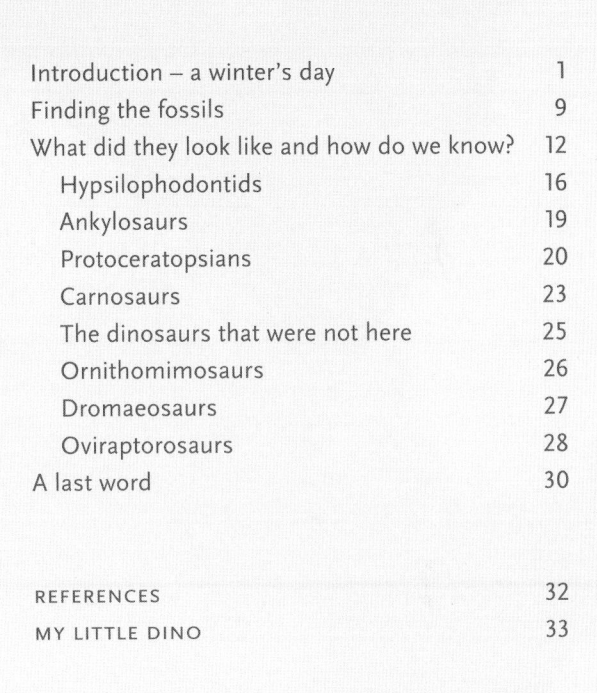

Introduction – a winter's day	1
Finding the fossils	9
What did they look like and how do we know?	12
Hypsilophodontids	16
Ankylosaurs	19
Protoceratopsians	20
Carnosaurs	23
The dinosaurs that were not here	25
Ornithomimosaurs	26
Dromaeosaurs	27
Oviraptorosaurs	28
A last word	30
REFERENCES	32
MY LITTLE DINO	33

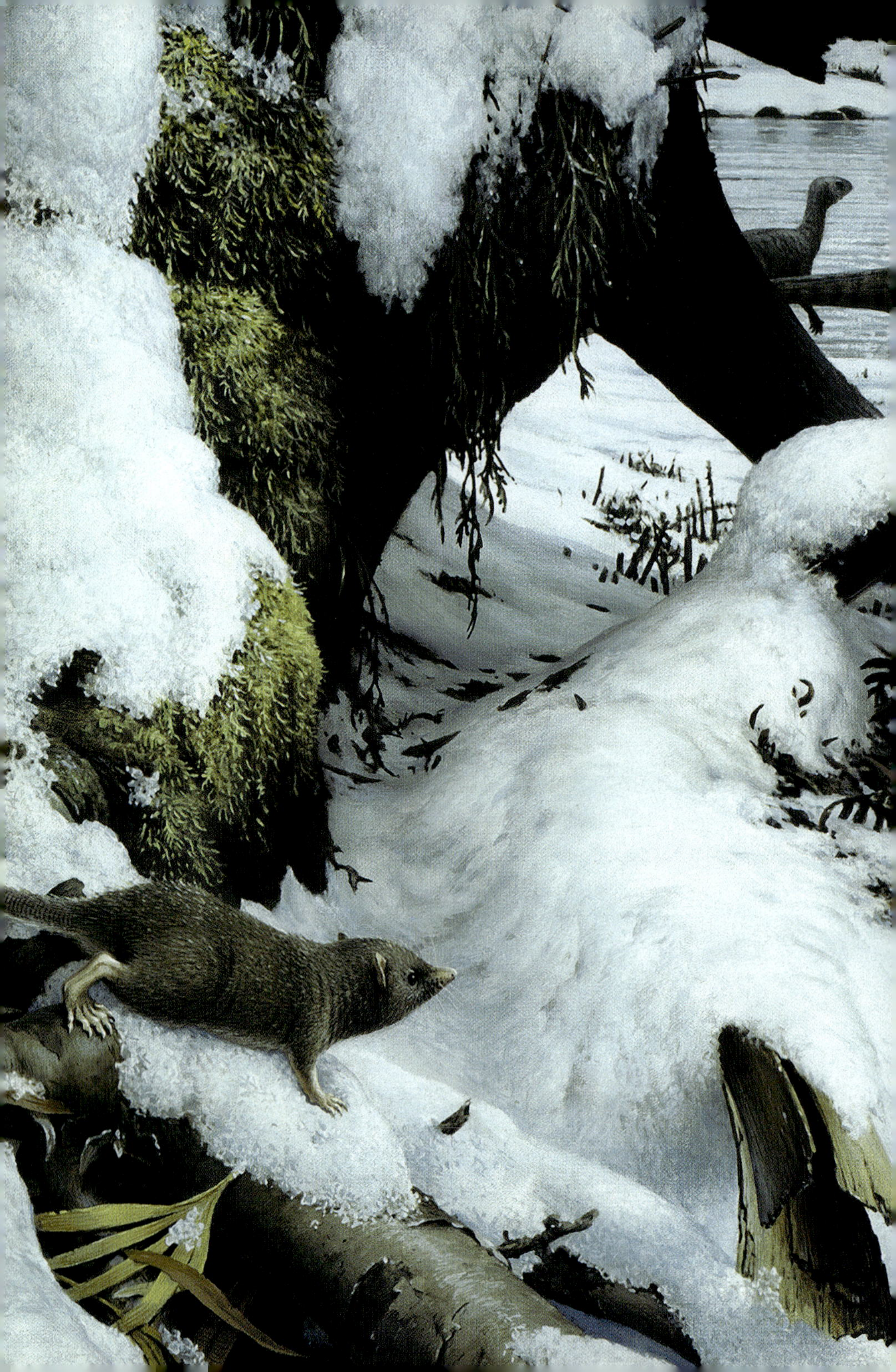

A winter's day

Imagine a winter's day 106 million years ago on what is now the coast of Victoria. Except for the Aurora Australis, or Southern Lights, and the moon, it is dark all day. No ocean was visible in the soft, pale light; rather, a broad flood plain stretched far to the south. You could walk all the way to Antarctica if you wanted to. For at that time, Australia was not where it is now but was nestled against Antarctica which straddled the South Pole.

An ausktribosphenid (a possible placental mammal) peers at a group of hypsilophodontid dinosaurs crossing a frozen lake near what is now Inverloch, Victoria, during a polar winter.
ARTIST: PETER TRUSLER

The dinosaurs that lived in what is now coastal Victoria dwelled within the Antarctic Circle. This meant that each winter the area they inhabited experienced months when the sun did not shine. By the same token, in summer there were months when it did not set at all.

At that time, Victoria was one of the coldest places on Earth. Nowhere else at any time when the dinosaurs lived was it cold enough to form ground that remained frozen all year around—permafrost—like that which exists now in northern Alaska. But that was the case when dinosaurs lived in Victoria. Despite the cold, the vegetation was lush. Shortly after the Victorian dinosaurs lived and died, the flowering plants exploded onto the scene. But because they were extremely rare when the Victorian dinosaurs were present, it was a green world. While some of these were small plants, others took the form of trees, perhaps as tall as 10 metres high, although it was a polar environment. On the tundra of northern Alaska today, the 'trees' are only a few centimetres high. Scientists are still trying to work out why trees were able to survive in these polar environments.

Beneath the trees grew an understory of bushes. Around them skirted a variety of dinosaurs. Most common were the plant-eating hypsilophodontids which came in a variety of sizes and shapes. These dinosaurs thrived in this habitat close to the South Pole. They were active during the winter darkness in this harsh environment and developed structures on their brains which enabled them to see in low light conditions so they could continue to feed efficiently at night. They did not hibernate or sleep for prolonged periods through the winter. Rather, like musk oxen today, they were able to eke out a living from the plants amongst which they lived during the winter, even if they were buried in snow. By feeding during the winter, they avoided predators at that time of year because the carnivorous dinosaurs hibernated. Today, when

A group of hypsilophodontids are active under the Aurora Australis during the polar winter in Victoria 110 million years ago, while a carnivorous ornithomimosaur hibernates.
ARTIST: PETER TRUSLER

Opposite page: *Leptoceratops*, from Alberta, Canada. The Victorian *Serendipaceratops* was very similar in appearance.
ARTIST: PETER TRUSLER

it is cold, the only land animals with backbones that are active are warm blooded. At present these are the mammals and birds. Cold-blooded reptiles are not active at such times. Because of this, we think it is almost certain that *Leaellynasaura*, and all other hypsilophodontid dinosaurs, for that matter, were warm blooded. Today in Fairbanks, Alaska, which is about as cold as Victoria was 110 million years ago, the only vertebrates one sees in winter are the warm-blooded birds and mammals.

During the warmer times of the year, at least three different kinds of small, agile, carnivorous dinosaurs lurked in the bushes, patiently waiting

An allosaurid carnosaur (centre) attacks *Muttaburrasaurus*, a large ornithischian (right), while *Leaellynasaura*, a small hypsilophodontid, dashes to safety (detail).
ARTIST: PETER TRUSLER

to make a meal of the similar sized hypsilophodontids. Thus it would seem to have been a very great advantage for a small, plant-eating dinosaur living close to the South Pole to be active during the winter, and so escape predation for part of the year.

Less common are the fossils of a number of different larger dinosaurs. These include armoured dinosaurs (ankylosaurs); protoceratopsians (pre-horned dinosaurs); large, fierce predatory carnivores; agile, emu-like ornithomimosaurs which were omnivorous, eating both plants and animals; and the oviraptorosaurs with their peculiar skulls. Except for the ankylosaurs,

POLAR DINOSAURS OF AUSTRALIA

The ankylosaur *Minmi*, known from two skeletons from Queensland, is the most complete dinosaur from Australia. A few isolated teeth, some pieces of body armour and a vertebra of an ankylosaur that was about the same size as *Minmi* are known from Victoria (detail).
ARTIST: PETER TRUSLER

all of these rare dinosaurs are unknown elsewhere in Australia. We only know these dinosaurs were present because the rare individual bones of these animals happen to be very distinctive, enabling scientists to assign them to a specific group. On the other hand, many of the fossil bones that have been collected from those sites cannot be identified more precisely than to say it is a bone of a fish, or that it is the bone of a land animal.

Fish and turtles lived in the streams running across the flood plain as you might expect. For the most part both the turtles and fish were quite primitive forms, only distantly related to their modern counterparts. One of these is the oldest known example of the horned turtles or meiolanids. What you would not expect to find was a giant amphibian, a very distant relative of frogs and salamanders, that bore a closer resemblance in size and shape to crocodilians. Yet it was not related to crocodiles. Rather it seems that crocodiles replaced them. When this amphibian *Koolasuchus* was alive, it was a 'living fossil' because all the other species of this kind of animal were millions and millions of years older than the Victorian one.

Victoria is not the only place on Earth where the fossils of dinosaurs are known from sites that were once close to one of the poles. Apart from a site in New Zealand, all the other ones are still close to a pole. Because of that, just getting to them is a major operation, as is digging fossils once there. It is because of the far greater costs of collecting from these other sites, that of all the polar dinosaur sites discovered to date, we know more about the Victorian ones than any of the others. This will probably not always be so, because at some of these other sites, dinosaur bones are far more common than in Victoria. If these sites are excavated to the same extent as the Victorian ones, then the amount of fossils collected will no doubt far outnumber those from Victoria, because once on the site the fossils are far easier to excavate and there are more of them.

The crocodile-like amphibian *Koolasuchus* lurks, awaiting hapless prey.
ARTIST: PETER TRUSLER

POLAR DINOSAURS OF AUSTRALIA

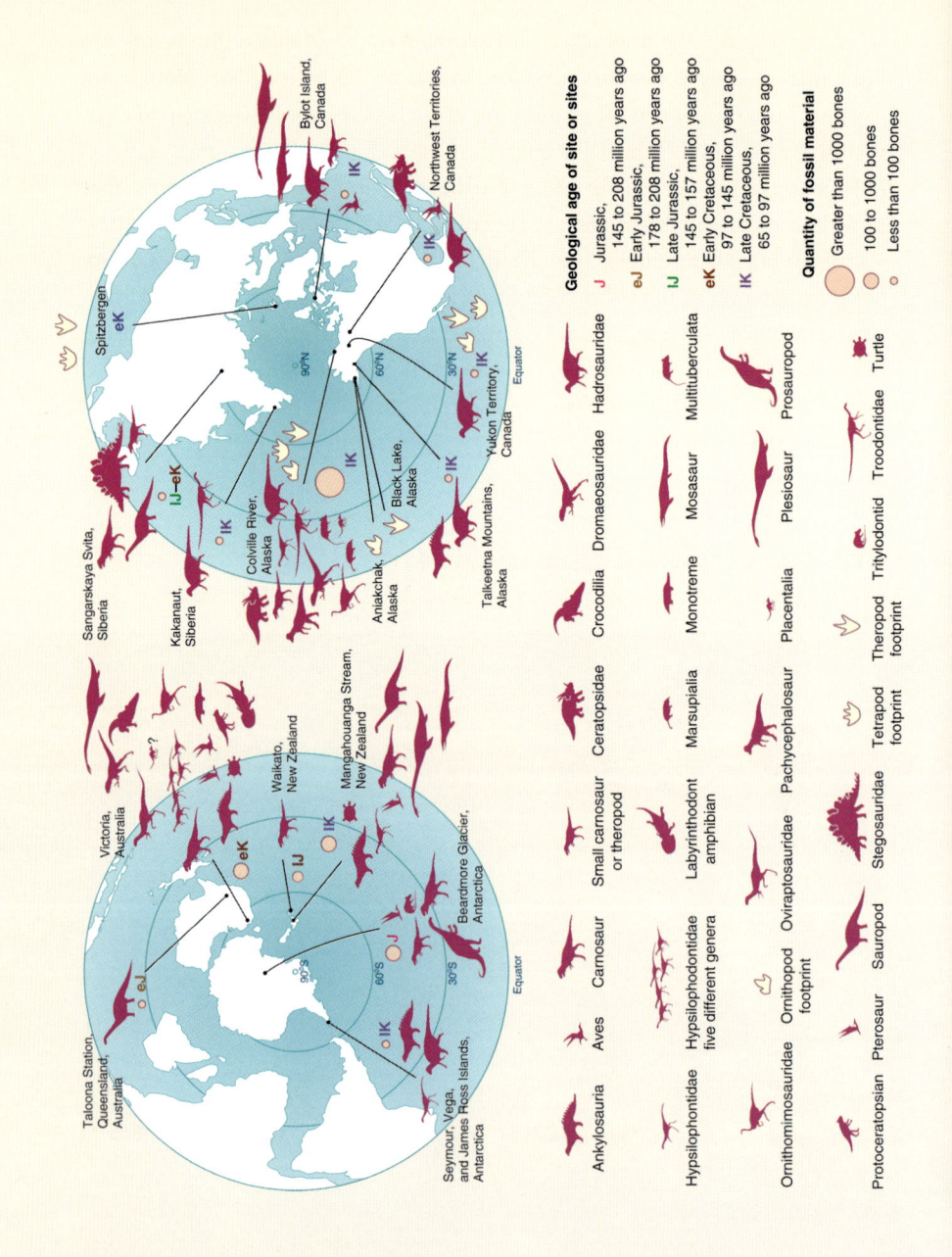

Sites of Jurassic (208 to 145 million years ago) and Cretaceous (145 to 65 million years ago) polar dinosaur assemblages. The base map shows the continental configuration in the Late Early Cretaceous (100 million years ago)

Finding the fossils

On 7 May 1903, a geologist by the name of William Hamilton Ferguson was walking along the rocky shore platform near Inverloch, Victoria. He was looking for coal deposits. On that day he did not find much coal. What he did find, however, was Australia's very first dinosaur bone. It was a 52 mm long claw of a meat-eating dinosaur. For the next seventy-five years it would not only be the first dinosaur bone found in Victoria, but also the last.

The first dinosaur from Australia to be discovered. Found near Inverloch, Victoria. PHOTO: JON AUGIER

That was to change in 1978 when three people—Rob Glenie, Timothy Flannery and John Long —returned to the site where Ferguson found his claw. John found another bone almost as soon as they reached the site. Over the next several months, Tim returned frequently to look for dinosaur bones on the shore platforms between Inverloch and San Remo. As he walked along that shore, hour after hour, he carefully examined all the exposed rock. He confined himself to the shore platform where the rock was well exposed so that he had a good chance to see any fossil bones that were present.

Realising that the same kinds of rocks occurred on the shore platforms of the Otway Ranges, Victoria, Tim and others searched that area in the same way in 1979 and 1980. As a result, a site was found which was later officially named Dinosaur Cove.

Left: Dinosaur Cove, Victoria. **Right:** Excavating a tunnel at Dinosaur Cove. The space where the men are standing was blasted out with explosives. They are digging down to the 10 cm thick fossil layer which lies about 1 metre beneath their feet.

Dinosaur Cove was excavated in search of fossils every summer but one between 1984 and 1994. Getting to the site was the first of several difficult tasks. One had to hike down a steep track to reach the cove. While people could walk to the site, in order to transport the heavy equipment required to dig the fossils and remove tonnes of fossil-bearing rock from the cove, it was necessary to construct a flying fox. What is a flying fox? It is a steel cable along which a basket runs which can carry things from one place to another. This flying fox was built by a volunteer, one of many hundreds who worked on the site over the years. Once the equipment and the people to operate it were on the beach, tunnels were dug into the hard sandstone to get to the fossils. This involved a lot of hard, dirty work on the part of the volunteers but they were eager to find dinosaurs. Digging the tunnels involved using rock drills to put holes into the rock, into which explosives were then placed. Once the explosives were detonated, the volunteers had to shovel out all the broken rock. The tunnel was cut with explosives placed 1 metre above where the dinosaur bones were thought to be so as not to damage them. After the tunnel was created, its floor had to be excavated to reach the fossil bones. The volunteers then carefully examined every fragment from the layer of rock where dinosaur bones were known to occur. Sometimes fossil bones were present and sometimes not. A lot of effort went into digging tunnels that sometimes didn't have any good fossils in them.

Left: The Flat Rocks site near Inverloch, Victoria. **Right:** The daily ritual of removing the sand from the Flat Rocks site that had accumulated since the last high tide.

At the end of the 1994 dig at Dinosaur Cove, there was nothing left to excavate. By then, half a dozen different dinosaurs had been found at that site.

In the meantime, another dinosaur site named Flat Rocks had been found only a few kilometres from where Ferguson had collected the first Australian dinosaur bone. This site is near Inverloch and continues to yield dinosaurs even to this day. Tunnelling is not necessary there for the site is on the open shore platform. Despite this advantage, it has other difficulties that make it a lot of work for the volunteers excavating it. The main problem is that the hole where the fossils are collected is covered each high tide, and the tide carries sand that fills the depression made by the dinosaur diggers.

A few dinosaur bones have been found at each of about a dozen other sites along the Victorian coast line. However, none of these other sites have yielded enough fossils to make extensive digging worthwhile.

Fossils being collected from the Liscomb Bonebed in northern Alaska. This is just one of the numerous polar dinosaur sites known from the banks of the Colville River.

What did they look like and how do we know?

Most people who are curious about dinosaurs have an idea of what they looked like when they lived. Those pictures are in people's minds because ever since the first dinosaurs were found in the nineteenth century, artists have been attempting to draw, paint or even sculpt them as living animals. The dinosaurs of Victoria are no exception. Artists, particularly Peter Trusler, have recreated these dinosaurs in a polar setting. Some of these images have been commissioned for use in books and for supporting documentaries. Others have found their way onto stamps. An Australia Post stamp issue in 1993, at the time of the release of the movie *Jurassic Park*, focused on Victorian dinosaurs.

Two partial skeletons of the hypsilophodontid *Leaellynasaura amicagraphica* are known from Victoria. All the other Victorian dinosaur specimens are single bones, jaws, or teeth. For this reason, if all we knew about dinosaurs were the specimens from Victoria, we would have only the slightest idea of what they looked like. Fortunately, however, there are many dinosaur skeletons known throughout the world. Most of these single bones from Victoria are almost identical to a bone from a complete skeleton from somewhere else. We can work out, therefore, what the whole skeleton and bodies of Victorian dinosaurs probably looked like. Imagine if all red kangaroos were extinct and all you had was a single fossil thigh bone of one of these animals. If you also had a complete skeleton of a grey kangaroo, you could make a reasonable guess at what the red kangaroo skeleton looked like by comparing the proportions of the thigh bones of these two animals, using the various grey kangaroo bones as models of the corresponding ones of the red kangaroo.

In the best circumstances, the artist and the scientists work closely together to come up with an accurately reconstructed scene in which dinosaurs are shown in their life positions. The job begins with the skeleton itself. That is

Reconstruction of the hypsilophodontid dinosaur *Leaellynasaura* as a corpse.
ARTIST PETER TRUSLER

all we know directly about each dinosaur. Over that the muscles are painted as they probably would have appeared. Although birds and crocodiles look quite different, when you compare the different muscle masses inside them, you find that most of the same muscles are present and these attach to similar places on the bones as the dinosaur muscles did. In some animals, a muscle is big while in another animal, the same muscle is small and perhaps others are bigger. This is true of all the living birds and reptiles that we can examine. So, it is reasonable to assume that dinosaurs had the same muscles. The only question is how big were they and where exactly did they sit on the skeleton. This the artist works out by carefully examining the skeleton.

Reconstruction of the skeleton and musculature of *Leaellynasaura*. This is an important intermediate step in reconstructing this dinosaur as it would have appeared when a living animal.
ARTIST: PETER TRUSLER

13

POLAR DINOSAURS OF AUSTRALIA

An Australia Post stamp issue commemorating the dinosaurs of this country.
ARTIST: PETER TRUSLER

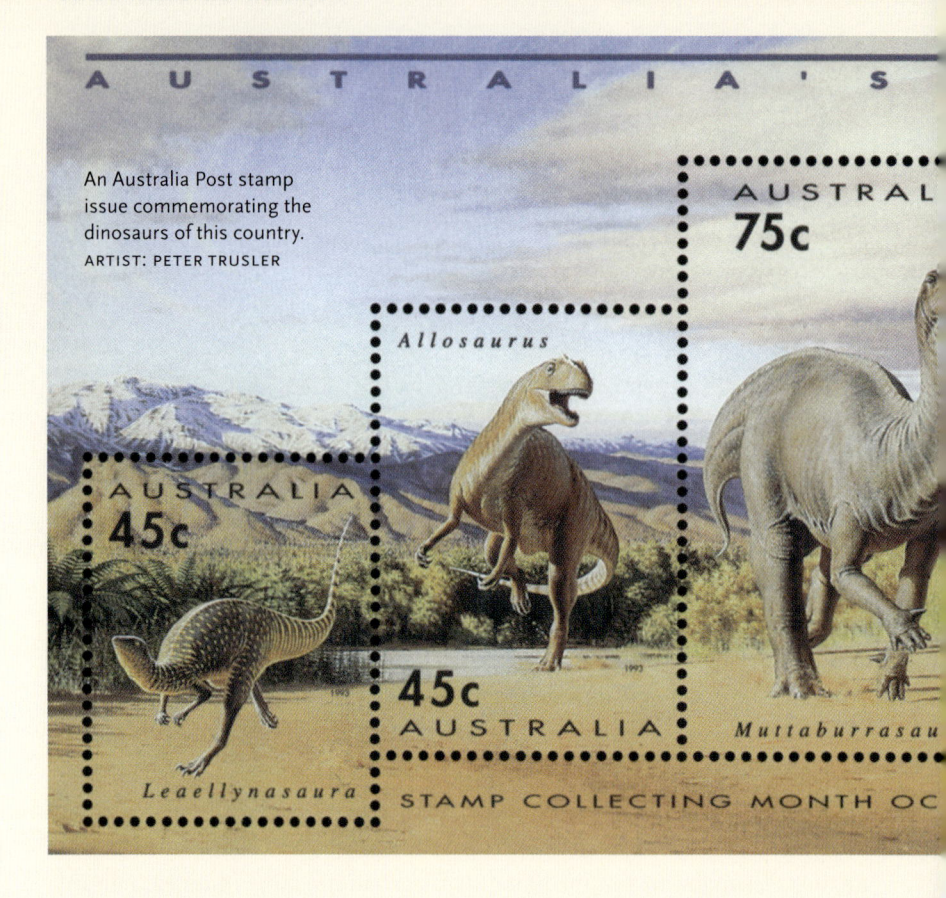

Once the meat is on the bones (muscles are the meat on a skeleton), the artist covers the dinosaur with skin. The next question is what colour was the animal? There is almost never any sign of the colour of the dinosaur preserved. So what the artist does is to pick the living animal which he or she thinks was most similar to the dinosaur and uses the colour scheme of that animal as a guide.

The dinosaur did not live on a blank stage. So the next step is to reconstruct the environment in which the dinosaur existed. Did the animal live on a mountain top, a desert, in the middle of a vast plain or some other environment? Geologists studying the rocks in which the dinosaur was found are able to determine its habitat. The vegetation in which the animals lived can be restored by examination of the plant fossils found in the same rocks with the dinosaurs.

With all this information in mind, the artist first sketches a scene and then, after double-checking with the palaeontologists who are studying the dinosaurs and plants and the geologists who reconstructed the environment, begins to paint a final version of it. By this time the artist has in mind not only the time of year, but also the time of day or night when the scene takes place, the angle of the sun light or moon light. This is very important for determining the colour of the sky. The artist also needs to think about the background. What sort of plants, leaf litter or snow, for example, were present. If there is a lake or pond, should there be ice on it? What about mist in the air? Finally, the artist needs to consider what the dinosaurs are doing. Even so, as the painting evolves, questions come up that need to be asked of the scientists in order to make the painting more accurate.

The end result is a recreation of a moment in the past when the dinosaurs lived. Such a picture is not only a pleasant thing to look at but is, in a very real sense, a scientific theory.

Hypsilophodontids

Half the dinosaur species known from Victoria belong to this one family, the Hypsilophodontidae, and far more than half the dinosaur specimens found in this state belong to that one family. Hypsilophodontids were small, agile plant eaters, perhaps the potoroo equivalent of the dinosaur world. Obviously, when the dinosaurs lived, Victoria had something about it that this group liked very much. As explained previously, there is good evidence that they were capable of seeing during the dark polar winter, did not hibernate, and were warm blooded.

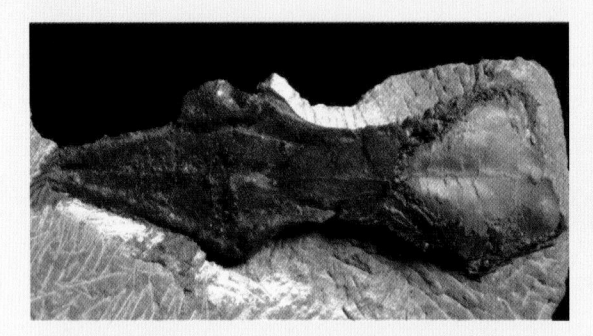

Looking down on the top of the 52 mm long skull of *Leaellynasaura*. An impression of part of the brain is visible on the right side of the skull.
PHOTO: STEVE MORTON

The best known of the Victorian hypsilophodontids is *Leaellynasaura*, for we have parts of her skeleton, not just single bones and teeth like all the other dinosaurs from this state. She would have stood no more than 40 cm high. Like all hypsilophodontids she had a body shape similar to that of a kangaroo; but she did not hop, have hair or external ears. Like the kangaroo, she was a plant eater.

Since the mid-1990s, many dinosaurs have been found in Liaoning Province, China, that have feathers on them. Feathers would have provided excellent insulation to keep the polar hypsilophodontid dinosaurs of Victoria warm. However, all the dinosaurs found so far with feathers are theropods, the group of carnivorous dinosaurs. Hypsilophodontids are not theropods. The skin of the closest relatives of hypsilophodontids we have a fossil record of is more like that of a lizard, completely lacking feathers or hair. There is another way in which the hypsilophodontids could have insulated themselves against the cold. They could have built up layers of body fat. This idea cannot be tested because fat does not fossilise, but it is the only way known to insulate an animal if it does not have feathers or fur, so it probably is what the cold-loving hypsilophodontids did.

To name a dinosaur, the first thing you need is a dinosaur specimen that is different to any other dinosaur. Next you have to check and be sure that the name you plan to call your dinosaur has not already been used. If it hasn't, then you can name it following a set of international rules for doing so. The name can be for some feature of the dinosaur. *Megalosaurus*, for example, means simply 'Big Reptile', which it certainly is. Or a dinosaur can be named for the place it comes from or to honour someone. *Leaellynasaura* was named for a little girl, Leaellyn Rich (the name means 'Leaellyn's Reptile'). Besides *Leaellynasaura*, three other hypsilophodontid dinosaurs have been identified from Victoria. *Fulgurotherium* was named in honour of Lightning Ridge, New South Wales, where the first specimen was found. Its name means 'Lightning Beast' in reference to Lightning Ridge, and it was later found in Victoria. *Atlascopcosaurus* was named for Atlas Copco, a Swedish mining equipment company that supplied all the equipment to excavate dinosaurs from Dinosaur Cove. Had it not been for that company, all the dinosaurs from Dinosaur

Eight different kinds of femora or thigh bones from eight different species of hypsilophodontids. Each species is represented by a pair of thigh bones.
PHOTO: JON AUGIER

Two individuals of the hypsilophodontid *Atlascopcosaurus* wander along the bank of a river swollen by spring floods, near what is now Cape Otway, Victoria.
ARTIST: PETER TRUSLER

Cove would still be in the ground, so they certainly deserved to have one named in their honour. *Qantassaurus* was named in honour of Qantas, which air-freighted an exhibition about Victoria's dinosaurs called 'Dinosaurs of Darkness' all around Australia and to other parts of the world.

These various hypsilophodontid dinosaurs generally looked alike, as do the various species of kangaroos alive today. But they had individual differences in their bones and teeth that make us realise that they are quite different species. *Qantassaurus*, for example, had a very short, deep jaw which differed from that of all other hypsilophodontids. It was a sort of bulldog of the group. Besides the four Victorian species of hypsilophodontids that have been officially named, there are about five more found in Victoria that are still being studied, so are not yet ready to be named. We realise they are different from the named Victorian hypsilophodontids and from each other, but the specimens are so imperfect that until more complete fossils are found they won't be able to be named.

Ankylosaurs

The most complete Australian dinosaur, *Minmi*, belongs to the ankylosaurs. Two nearly complete skeletons of it have been found in eastern Queensland. For an ankylosaur, it is rather small, about the size of a bullock. Like other ankylosaurs, it had armour on its back. It also had armour on its belly which is most unusual, especially as these dinosaurs were mild mannered animals that would not have attacked any other animals unless provoked.

The skeletal evidence for ankylosaurs in Victoria consists of merely a few isolated teeth, two vertebrae or back bones, ribs, and parts of its armour in the form of small bones that were buried just beneath the surface of the skin. (If you cut an ankylosaur rib in two, its shape looks like the letter 'L'. This is unique to the ankylosaurs and one of the best ways of identifying them.)

From the few bits of the Victorian ankylosaurs that we have, it appears that one of these animals was about the size of *Minmi*, which makes it a small ankylosaur indeed. The structure of its teeth are quite different from *Minmi*, however, so it seems that a different ankylosaur lived in Victoria. More of this animal needs to be found before we can be sure. More surprising, however,

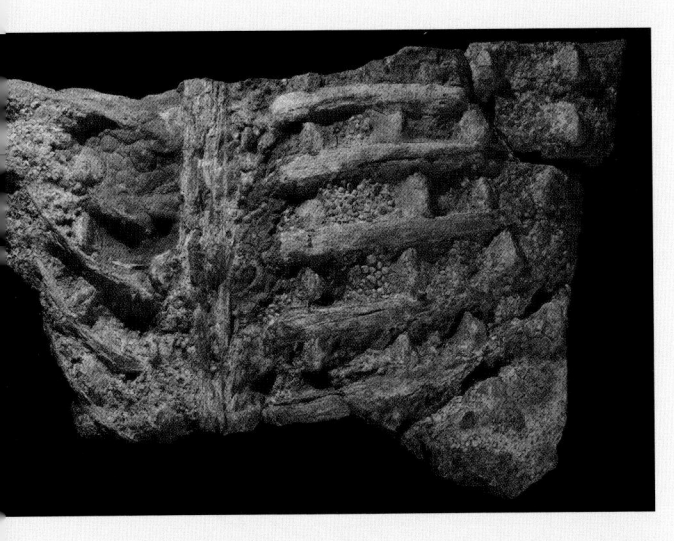

Left: Part of the skeleton of the Queensland ankylosaur *Minmi*. An ankylosaur similar to *Minmi*, if not *Minmi* itself, lived in Victoria. PHOTO: FRANK COFFA
Right: A tooth of the Victorian ankylosaur. PHOTO: STEVE MORTON

is the nature of what is probably a second Victorian ankylosaur. While *Minmi* and the first Victorian ankylosaur mentioned were small ankylosaurs, this second Victorian one was really tiny. The length of its dorsal vertebra was only about one-third that of *Minmi* or the first Victorian ankylosaur. It would have been no larger than a wombat if even that big. Why such a small ankylosaur should have lived in Victoria is not clear. However, the evidence for the existence of this tiny ankylosaur is based on a single vertebra. So while Victoria may have been the home of the smallest ankylosaur that ever lived anywhere on Earth, it is not yet certain that this is so.

If you look at a map of the world as it was 120 million years ago, you would see the continents in quite different positions compared to today. You would see a link from Australia via Antarctica to South America, and thence to the Northern Hemisphere. That being the case, it is surprising to learn that while ankylosaurs are known from Antarctica and South America, they occur only from rocks more than forty million years younger than those found in Australia. Did the ancestors of *Minmi* and the Victorian ankylosaurs use that route to enter Australia? Or was it from the much older Australian stock that the generally much younger ankylosaurs on those other continents arose? The gap between Australia and Asia was much greater then than now so that route does not seem too likely. So how did they reach Australia? That is a mystery waiting for a solution!

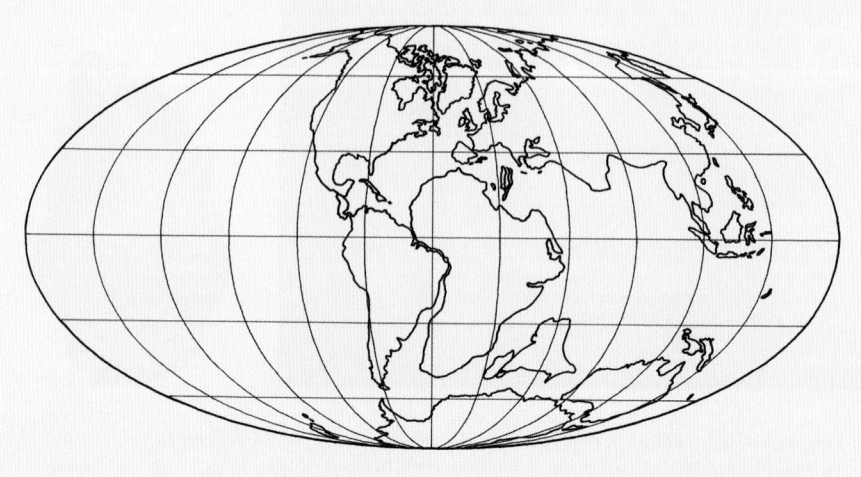

Map of the world 120; million years ago. Note Australia's position close to the South Pole.

Protoceratopsians

The protoceratopsians were the ancestors of the horned dinosaurs or ceratopsians, which include the well known *Triceratops*. Unlike the ceratopsians, the protoceratopsians were much smaller animals that lacked horns and could be considered to be the 'sheep' of the dinosaur world. Until recently, protoceratopsians were well known from North America and Asia only from rocks millions of years younger than the one specimen from here in Australia. Because the rocks here are millions of years older than those in North America and Asia, where so many protoceratopsians have been found, we never expected to find any trace of them in Australia. Imagine our surprise when a fossil was found that seemed to belong to that group.

The Victorian fossil was from the forearm of one of these dinosaurs. In particular, it is the same bone, the ulna, which your elbow is a part of. We could tell it was a protoceratopsian because this bone has a very unique shape

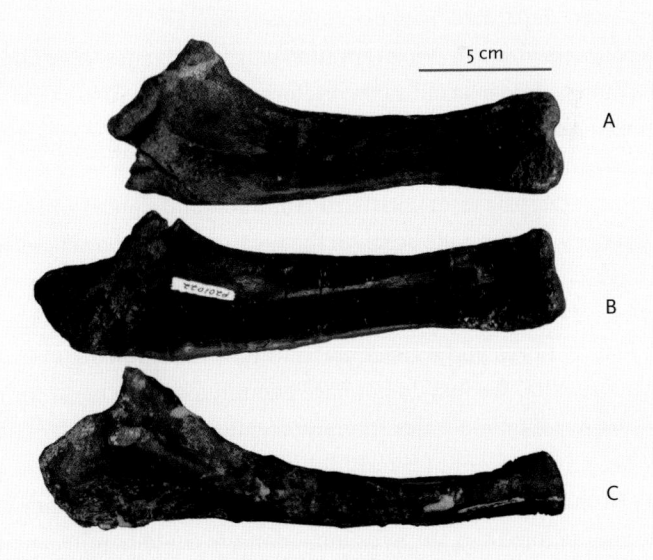

Comparison of the ulnae of two protoceratopsians (A & B) and one theropod dinosaur. Although therapods have very reduced ulnae, they are quite unlike protoceratopsians. A. *Serendipaceratops* from near Kilcunda, Victoria. B. *Leptoceratops* from Alberta, Canada. C. Theropod from Victoria. PHOTO: STEVE MORTON

not found in any other dinosaur, or any other animal for that matter. It is flattened from side-to-side and thick from top to bottom. Most ulnas are more like a long rod. However, because there is only one bone and that is from the forearm, some palaeontologists still have doubts about it. Bones of the forearm are rarely distinctive enough to be identified by themselves. For this reason most palaeontologists have difficulty telling what kind of animal they belong to because they rarely have to identify isolated ulnae. So because of this, and the fact that the idea of a protoceratopsian in Australia seems so outlandish, many palaeontologists still do not think this fossil could belong to that kind of dinosaur. Owing to these misgivings, finding more of this dinosaur is a major goal of our regular fossil digs here in Victoria.

The Victorian specimen is not only similar to protoceratopsians generally, but it is so similar to one of them in particular, called *Leptoceratops*, that we have a very good idea of what the whole animal looked like. This is because entire skeletons of *Leptoceratops* are known. We can assume that the whole Victorian protoceratopsian may have looked much like *Leptoceratops* (see page 2). Despite looking so much alike, these two dinosaurs lived fifty million years apart and were separated by 15 000 kilometres.

Several years after the Victorian ulna was found, a few specimens of protoceratopsians, almost as old as the Victorian one, were discovered in eastern and western Asia. So the Victorian specimen is no longer considered unusually old for the group.

Did the ceratopsians originate in Australia and spread to the Northern Hemisphere? With the wide ocean to the north of Australia separating it from Asia it seems incredible that this could be possible. But somehow they had to get between the Northern Hemisphere and Australia so it is at least as likely that they started here and walked north, versus first evolving elsewhere then reaching Australia. We just do not have enough fossils to know for sure.

Like the hypsilophodontids, the protoceratopsians were plant eaters with a horny beak at the front of the skull, useful for nipping off vegetation. But the beak of the protoceratopsians was stronger. Presumably they could deal with tougher plants than the hypsilophodontids could handle, such as cycad fruits.

After having the Victorian protoceratopsian specimen in Museum Victoria for more than ten years and not having found any more of the same species, we decided to name it. Pat Vickers-Rich and I called it *Serendipaceratops* which means 'the fortuitously discovered horned dinosaur', even though like all protoceratopsians, it did not have a horn. But the lack of a horn is no problem with a scientific name. For example, *Protoceratops* does not have a horn either, even though its name means 'first horned'. Added to the generic name was the species name *arthurcclarkei*. In the case of *Tyrannosaurus rex*, *Tyrannosaurus* is the generic name and *rex* is the species name. The species is a subdivision of a genus so there can be more than one species in a genus. For example there is another species of *Tyrannosaurus*, it is *Tyrannosaurus turpanensis*. In the case of *Serendipaceratops arthurcclarkei*, the species name is in honour of renowned science fiction writer, Sir Arthur C. Clarke. His interest in science was first sparked by learning about dinosaurs. He lives in Sri Lanka, an ancient name of which is Serendip, so the combination of the generic and specific names is most apt.

Carnosaurs

All was not idyllic in polar Victoria 110 million years ago. There were large carnivorous dinosaurs called carnosaurs roaming the land. One of them is known from an ankle bone about the size of an adult's clenched fist. That small part of the skeleton is enough to tell that it is related to the well-known North American carnivorous dinosaur *Allosaurus*. It is probably not *Allosaurus* itself but another carnosaur in the same family. If that is the case, this dinosaur when alive was a living relict because all the other allosaurids had died out more than twenty million years earlier. From the size of that one ankle bone, it is possible to work out the approximate size of the whole animal. This is possible because whole skeletons of *Allosaurus* exist that have that bone in place. On that basis, the Victorian allosaurid was 3 to 4 metres high when it was standing upright on its hind legs.

Another carnosaur is known from just ten per cent of a claw. That small fragment is just enough to show that while it is a carnosaur, it is not an allosaurid. So this means that at least two large carnivorous dinosaurs were present in Victoria. However, the two specimens we have were of animals that lived ten million years apart so maybe there was only one large carnivorous dinosaur alive here at each time. As there are so few specimens, we really

The one fossil footprint of a stegosaur known from Australia, specifically from Broome, Western Australia.
PHOTO: JOHN LONG

cannot say how many such dinosaurs were around. All that we are safe in saying is 'at least two'.

Most of the plant-eating dinosaurs that lived in Victoria for which we have a fossil record are too small to have made much of a meal for these large carnosaurs. The only ones known from Victoria that were big enough to feed them are the armoured ankylosaurs. So the presence of these large carnosaurs, few as the specimens are, points to the probable presence of other large plant-eating dinosaurs here too. Unfortunately not a trace of these has yet been found in Victoria.

The dinosaurs that were not here

One of the 'missing' dinosaurs that might actually have lived in Victoria was the large plant eater *Muttaburrasaurus*, which occurred in Queensland. *Muttaburrasaurus* was a two-legged dinosaur that looked somewhat like an over-sized kangaroo. Another candidate could have been the stegosaur. *Stegosaurus*, perhaps the best known of dinosaurs, is characterised by the presence of large triangular plates along its back and spikes on its tail. They had become extinct on other continents long before the Victorian dinosaurs lived and died. However, a fossil footprint from Broome in rocks the same age as those that yield the Victorian dinosaurs has been identified as a stegosaur. So maybe they were present in Australia but have just not been found yet in Victoria.

Another group of plant-eating dinosaurs also known in Queensland are the long-necked sauropods such as the well-known *Brontosaurus* (more properly known as *Apatosaurus*). They have not been found in Victoria and are unlikely ever to be because the polar climate of Victoria might not have agreed with them. They were probably cold blooded whereas other dinosaurs may well have been warm blooded. Being cold blooded makes good sense for a sauropod. Because they had small heads, delicate teeth and huge bodies, it is difficult to imagine how they would have eaten enough to keep warm through digesting massive quantities of food. An animal that

The Early Cretaceous ornithomimosaur is one of the dinosaurs found at Dinosaur Cove, Victoria (detail).
ARTIST: PETER TRUSLER

is warm blooded has to eat about ten times as much food as one the same size that is cold blooded.

Ornithomimosaurs

No other dinosaur could run as swiftly as these two-legged, emu-like dinosaurs. It is the proportion of the hind legs that tells us this. The thigh bone or femur is much shorter than the bones in the lower part of the leg, exactly as is found in the emu, ostrich and other large ground birds which are fast runners. These dinosaurs were about the same size as those birds. Although ornithomimosaurs are theropod dinosaurs and thus related to forms like *Tyrannosaurus rex*, most of them did not have teeth (except for the most primitive ones). They may have been omnivores.

In Australia, ornithomimosaurs are only known from Victoria. The Victorian one is called *Timimus*. Two femora or thigh bones of this dinosaur

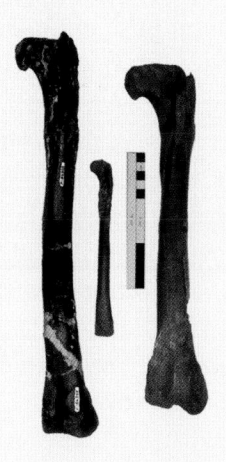

Left: Left, femur of an ornithomimosaur from the Late Cretaceous of Alberta, Canada. Centre, femur of a juvenile of *Timimus* from Dinosaur Cove, Victoria. Right, femur of an adult of *Timimus* from Dinosaur Cove, Victoria. **Right:** Isolated tooth of a dromaeosaur from Victoria.

were found within a few centimetres of one another. One is only about half the size of the other. Presumably, the smaller one is a juvenile of the same species. Yet despite their close proximity, it is quite unlikely that they were parent and offspring. This is because both bones were carried some distance by a flowing river after their skeletons had broken apart. It would be an incredible coincidence if two such different sized bones drifted together for perhaps kilometres before coming to rest next to each other. Compared with other ornithomimosaurs, the thigh bones of *Timimus* were particularly long and slender. It may well have been that *Timimus* was amongst the fastest of these speedy dinosaurs.

Dromaeosaurs

Probably the most agile and intelligent dinosaurs were the dromaeosaurs. These two-legged animals were active carnivores. Although much smaller than *Tyrannosaurus rex*, they had a brain that was proportionally much larger, about the size found in an emu or ostrich. Dromaeosaurs include the well-known *Velociraptor* from Mongolia.

The Australian record of this family consists only of isolated teeth. Of these, the vast majority have been found in Victoria. Almost all of the teeth found thus far were shed by the animal during the course of its lifetime. We can tell this because there are no roots preserved beneath the biting part or crowns of the teeth. What happens in animals like crocodiles that shed many teeth in the course of their life is that the roots are dissolved by the body of the animal, and without that support, the tooth eventually falls out of the jaw. If you get a chance to look at your own baby teeth, you will find there are no roots in them and it is exactly for the same reason. [Most dinosaurs and other reptiles used their teeth only to grasp their prey. They do not chew it up like we do. By replacing their teeth frequently, they always have sharp teeth. Why don't we replace our teeth all through life? Because our upper and lower teeth fit together very precisely. When we lose a lower premolar, the two upper premolars above it are unable to function until it is replaced. So it is advantageous for us to replace as few teeth as possible.]

Probably three different kinds of dromaeosaurids are represented amongst the Victorian specimens. They were all about the same size, probably between 1 and 1.5 metres tall. What distinguishes them is the teeth. Some are like a saw blade with large cutting serrations on both the front and the back of the tooth. In others, the serrations are small on both sides or just present on one side and absent on the other. These teeth tell us that this family existed and that there were at least three different kinds of them. Unfortunately, we know little more than that.

Oviraptorosaurs

Fossils of these dinosaurs are rare anywhere. But most are known from eastern Asia and western North America. In Victoria, they are especially rare, for here the evidence for their presence is a single back bone and a bone from the lower jaw. From that meagre evidence about all that can be said is that the Victorian oviraptorosaur was about the same size as the smaller ones known elsewhere from more complete fossils. That is, they would have stood less than 1 metre high. Although toothless, these dinosaurs are regarded as part of the theropods, a group which includes all carnivorous dinosaurs, including the much larger *Tyrannosaurus rex*. As is typical of theropods, these animals walked only on their hind legs and had an elongated tail. The skull was very bird-like with large holes in it and the bone is thin-walled. In keeping with the bird-like nature of the skull, there is a crest of bone on the head, very similar to that of a cassowary. Like those birds, the crest may well have been covered with a horny sheath. Perhaps the oviraptorosaurs used them to butt heads with one another like cassowaries.

The light build and shape of the skeleton suggests that oviraptorosaurs were swift, agile animals. The sharp claws would have aided them in attacking their prey. There is a ridge in the roof of the mouth that is similar to the 'egg tooth' found in snakes that eat eggs. The name 'Oviraptorosaur' in fact means 'egg robbing lizard', as the first fossils found of them were associated

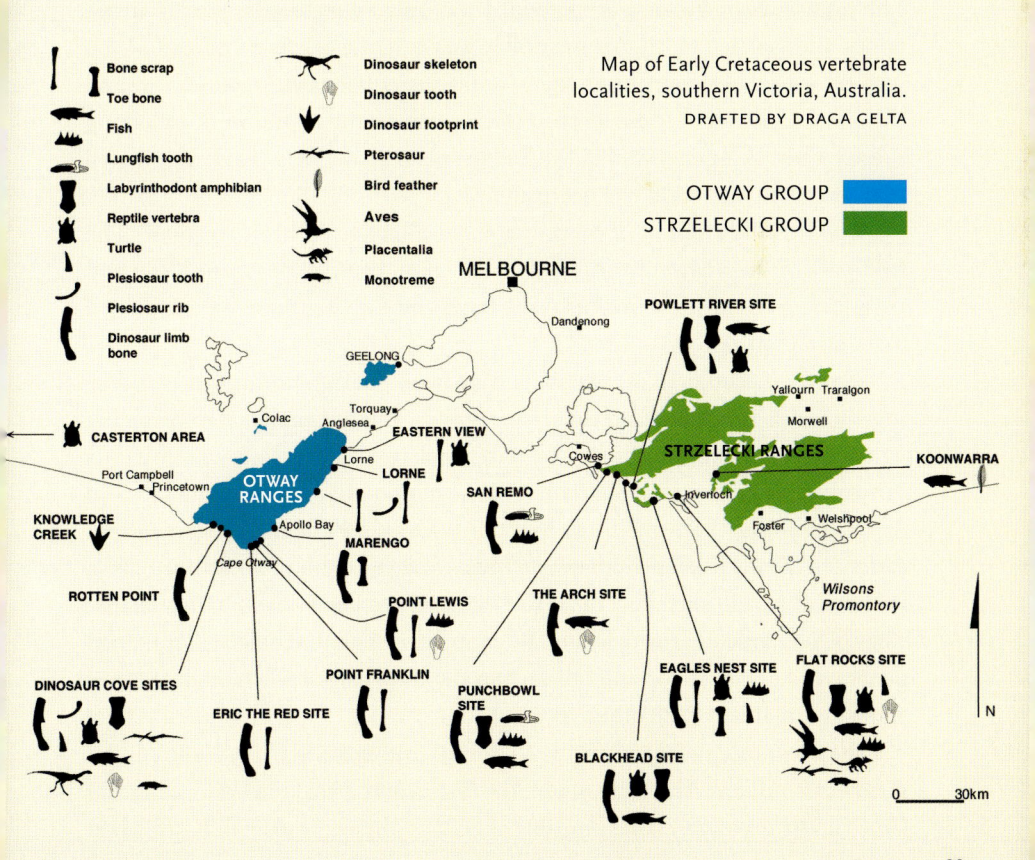

Map of Early Cretaceous vertebrate localities, southern Victoria, Australia.
DRAFTED BY DRAGA GELTA

with dinosaurs nests. Although scientists at first thought they were robbing the eggs, in recent years it was found that the dinosaurs were actually brooding their own eggs.

As with some of the other Victorian dinosaurs, the question of how they got between here and the other places they are known to occur is one that is not easily answered. In this case, the broad oceans then separating Australia from Asia and North America appeared to be a barrier these animals could not possibly have crossed. And yet they somehow did.

A last word

The dinosaurs of Victoria are known from only a few fossils, mostly single bones and isolated teeth. An incredible amount of work by hundreds of volunteers over a period of more than twenty years has been necessary to find this small collection. If one person were to have done all the work collecting these fossils, he or she would have had to spend about forty years, digging all year round, to find what has been achieved.

Why has the amount of dinosaur material collected been so small? The first part of the answer is that fossils are rare, even at Dinosaur Cove and Inverloch. Secondly, the rock is quite hard and digging therefore requires a lot of strenuous effort.

Why then should all of this effort have been put into finding and excavating Victoria's dinosaurs? In the first instance, there is no place in Australia where dinosaurs are common. The only other state with relatively numerous dinosaurs is Queensland; but there is almost no overlap between the kinds of dinosaurs found there and those found in Victoria. While the Victorian dinosaurs are mostly known from single bones and the teeth of small animals, a number of partial skeletons of large dinosaurs are known from Queensland. They belong to four named genera. By contrast, in Victoria there are six named genera and evidence for another six. So the dinosaurs known from the two states complement each other very well.

The most unique aspect of the Victorian dinosaurs is that they lived in a polar environment. The rocks where the fossils occur hold the evidence that at times it was extremely cold, probably as cold as anywhere else on Earth when

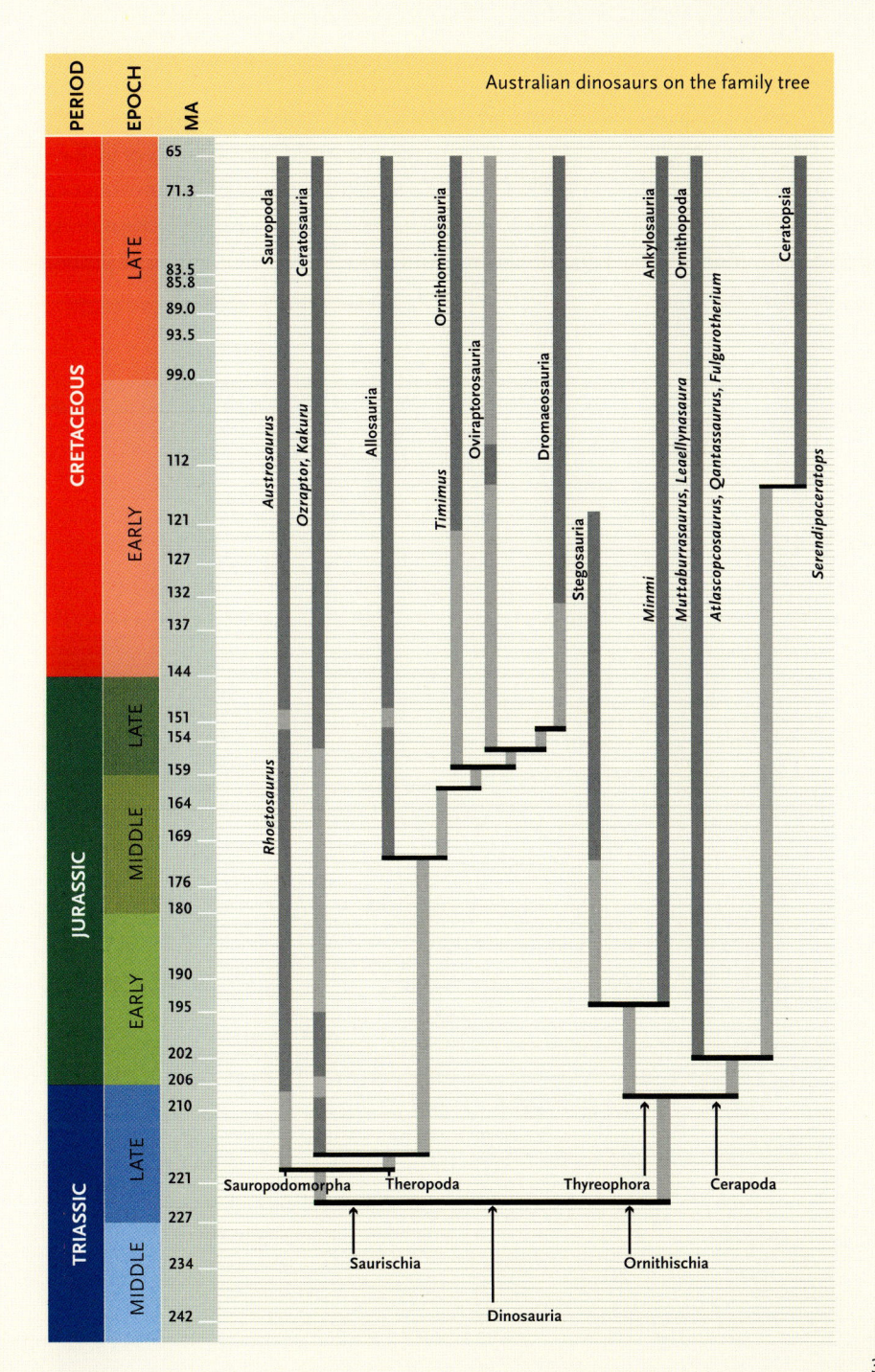

Australian dinosaurs on the family tree

the dinosaurs lived. In one hypsilophodontid, *Leaellynasaura amicagraphica*, we can see how the animal had adapted to these extreme conditions.

So Victoria contributes to our knowledge of Australian dinosaurs by having more species than any other state. That fact, coupled with their existence in a unique polar environment, makes these dinosaurs highly unusual. At present, more diverse polar dinosaurs are known from Victoria than any other place on Earth.

REFERENCES

If you want to learn more about Australia's polar dinosaurs, here are a few references to seek out.

Long, JA 1998, *Dinosaurs of Australia and New Zealand and other animals of the Mesozoic Era*, University of New South Wales Press, Sydney.

Rich, TH 1997, *Luck and persistence: a scientific career*, ABC website, viewed 25 September, 2006, http://www.abc.net.au/science/slab/rich/story.htm

Rich, TH & Vickers-Rich, P 2000 *Dinosaurs of darkness*, University of Indiana Press, Bloomington.

Rich, TH & Vickers-Rich, P 2003, *A century of Australian dinosaurs*, Queen Victoria Museum, Launceston and Monash Science Centre, Clayton.

Vickers-Rich, P & Rich, T 1993, 'Australia's Polar Dinosaurs', *Scientific American* [July]: 269 (1), July, pp 50–55. Updated and republished twice as Rich, TH, Vickers-Rich P, Chinsamy, A, Constantine, A & Flannery, TF 2000, 'Australia's Polar Dinosaurs', pp 323–30 in Paul, GS (ed.) *The scientific American book of dinosaurs*, St. Martin's Press, New York and Vickers-Rich, P & Rich, TH 2004, 'Dinosaurs of the Antarctic' pp. 40–7 in Rennie, J. (ed.) *Dinosaurs and Other Monsters. Scientific American Special, Scientific American* vol. 14, no. 2.

AUTHOR'S BIOGRAPHY

Thomas Rich is Curator of Vertebrate Palaeontology at Museum Victoria, Melbourne. He holds a PhD from Columbia University for work on fossil hedgehogs of the Northern Hemisphere. His main research interest is the early evolution of mammals, particularly those from the Cretaceous of Australia, South America and Africa. He has published two books on dinosaurs: *Dinosaurs of Darkness* and *A Century of Australian Dinosaurs*.

Leaellyn Rich is a solicitor at Freehills, Melbourne. She gained her BA and LLB from the University of Melbourne

ACKNOWLEDGEMENTS

The polar dinosaurs of Australia would be virtually unknown today were it not for two groups: first, the hundreds of volunteers who carried out the strenuous and often mind-numbing work that resulted in the collection of these fossils; secondly, those organisations that provided long term support for the project, particularly the Committee for the Research and Exploration of the National Geographic Society, Atlas Copco, and Orica.

The map on page 8 is reproduced with the permission of Indiana University Press. Complete citation: Thomas H. Rich and Patricia Vickers-Rich, *Dinosaurs of Darkness*, 2000, pp 14 & 185, Indiana University Press: Bloomington & Indianapolis.

'My Little Dino' was reprinted from the February 1995 issue of *Ranger Rick* magazine, with the permission of the publisher, the National Wildlife Federation.

My little dino

By Leaellyn Rich

MELBOURNE

DINOSAUR COVE

I always wanted my very own dinosaur. Then my parents found one for me—deep in an Australian mine!

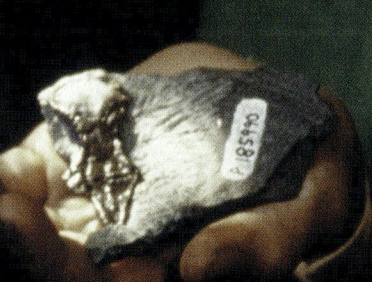

Here I am with the skull of a dinosaur my parents found. They named the big-eyed dino after me! The painting on page 39 shows what it may have looked like when it hatched. As an adult, it may have lived in groups, page 36.

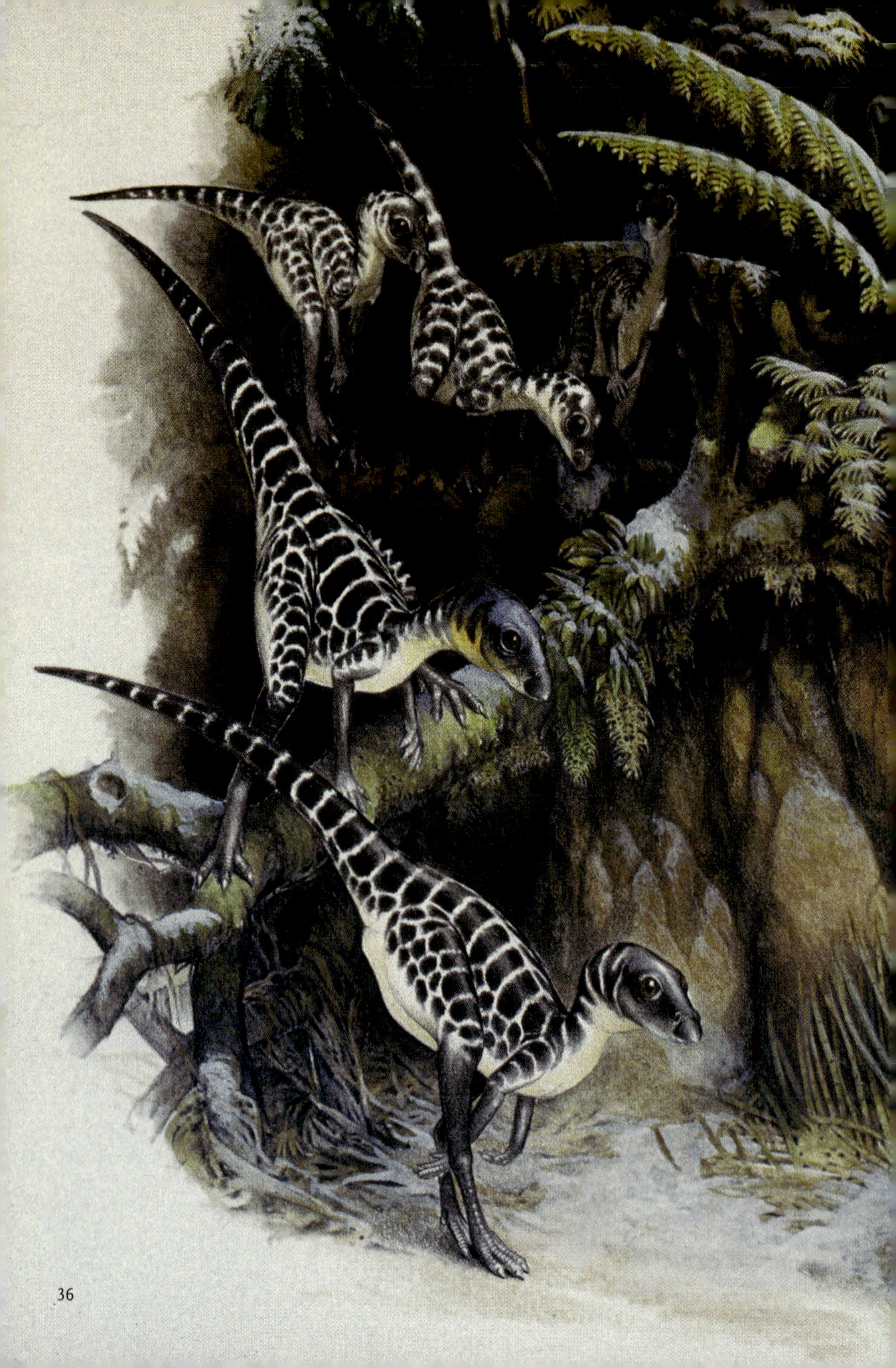

Leaellyn was asked by the editors of the children's magazine Ranger Rick *to write about the excavation at Dinosaur Cove from the unique perspective of a child who actually participated in it.* My little dino *was the result.*

When I was three, I had a book called *My Little Dinosaur*. It was about a boy who found a live dinosaur in a cave near his house. I started wanting a dinosaur too.

My dad worked with dinosaurs at a museum—so I asked him to get me one. 'Christmas would be a good time,' I told him.

Of course, Dad couldn't really bring me a dinosaur. And as I grew up, I forgot about wanting one. I was too busy with school, sports, ballet classes, and music lessons. But my dad remembered the dinosaur. And when I started grade five, something great happened.

Dinosaur bits and pieces

My parents work as palaeontologists (PAY-lee-un-TOL-uh-jists). That means they're scientists who study life from long ago. I've gone on heaps of fossil digs with them, starting when I was ten months old. In fact, I'd been working with them on a dinosaur dig that summer before I started grade five. But then school began, and I had to leave.

While I was at school, my dad called to say his volunteers had found a dinosaur skull. He wasn't sure what kind of dinosaur it was. But he thought it might be one that had never been found before.

To prove it, he'd have to see more parts of the skeleton. But those parts were probably scattered around and would be hard to find.

Luckily, my parents and the volunteers kept finding more bits of the dinosaur skeleton. Finally they had enough pieces to be sure. It *was* a new kind of dinosaur. And they named it after *me*!

DRAWING BY PETER SCHOUTEN

Big eyes for long nights

The new dinosaur was called *Leaellynasaura* (Lee-EL-in-uh-SOR-uh). It was little—about the size of a chicken. It had enormous eyes. And it used a bigger part of its brain for seeing than was normal.

My dad and mum think its big eyes and special brain may help prove that some dinosaurs were warm-blooded (able to keep their bodies warm in cold temperatures). Here's why: My dinosaur lived about 100 million years ago. Scientists think that way back then, Australia was attached to Antarctica. So the place where my dino was found would have been much closer to the South Pole than it is now.

In winter, it stays totally dark for three months of the year near the Pole. Big eyes are useful for seeing in the dark. So my dinosaur's big eyes and special brain would have been really useful in the winter—when the temperature was probably below freezing.

Dad and mum say that the eyes and brain seem to show that my dinosaur was active in the freezing winter. No cold-blooded animal living today could survive those temperatures. So my dinosaur was probably warm-blooded.

Dinosaur Cove

My dinosaur wasn't easy to find. It was buried deep inside a cliff along the coast near Melbourne. The cliff is beside a bay-like area called a *cove*. We called the area 'Dinosaur Cove'.

There, my parents found a layer of rock that they thought might have dinosaur fossils in it. So our family and some volunteers started digging near the bottom of the cliff.

At first, my parents and the volunteers dug into the rocky wall of the cliff with picks and large hammers. What a joke—that hardly cracked the rocks at all!

Then they tried using jackhammers and other power tools on the hard rocks. That worked a little better, but not well enough. Finally, they found some people who knew how to use explosives. They *blasted* their way in!

The explosives blew a huge hole in a layer of rock right above the dinosaur layer. The next step was to clear out the loose rocks and rubble. Then some volunteers used the jackhammers to dig down to the fossils. It was like working in a dinosaur *mine*!

I was too young to use a jackhammer. But I could help with the clearing out. And once the rocks were outside the mine, I got good at cracking them open to check for fossils. I also helped around the camp—cooking, cleaning up, and getting supplies.

ARTIST: PETER TRUSLER
THIS MATERIAL HAS BEEN REPRODUCED WITH PERMISSION OF THE AUSTRALIAN POSTAL CORPORATION. THE ORIGINAL WORK IS HELD IN THE NATIONAL PHILATELIC COLLECTION.

We spent nine summers digging at Dinosaur Cove. And we found many exciting fossils, including my dinosaur. I loved it there. The people were always really friendly. And there were plenty of cool animals to see—like colourful frogs and sometimes even fairy penguins.

I also liked helping out around the mine, though I didn't go into it very often. I've always been a bit uneasy in closed-in spaces. I used to dare myself to run right to the back of the mine. Then I would feel like a big hero if I could make myself stay there for more than two minutes at a time.

Even getting to the mine could be tricky. The only way was to climb down that steep, 90 metre cliff, holding onto a rope. Once you were at the bottom, you had to watch the tides. The sea is really rough there. During high tide, the water went into the mine. So when the tide started coming in, we had to get out of the way quickly—or get clobbered by a wave. I remember scrambling up the cliff lots of times just ahead of the water.

What a rush!

The work we did could be frustrating at times. I could sit for hours breaking up rocks without finding even a scrap of fossil bone. Other times I found only tiny, useless bone fragments.

But when I found a good fossil, I got such a rush of excitement. It was as if I'd made some strange connection with a totally forgotten being—a being that had lived millions of years before I was born, when the world was much different.

That is the most wonderful feeling. I got it no matter what kind of fossil I found. So you can just imagine how I felt when I first saw the fossils of my very own dinosaur! Thanks, Mum and Dad.

Here's Dad with the dino skull. He's standing just inside the mine. Dad and Mum figured out that the skull belonged to a new type of dinosaur. When alive, the dino was only about the size of a chicken.

PHOTOS: ©PETER MENZEL /WWW.MENZELPHOTO.COM